BEI GRIN MACHT SICH IHR WISSEN BEZAHLT

- Wir veröffentlichen Ihre Hausarbeit,
  Bachelor- und Masterarbeit

- Ihr eigenes eBook und Buch -
  weltweit in allen wichtigen Shops

- Verdienen Sie an jedem Verkauf

Jetzt bei www.GRIN.com hochladen
und kostenlos publizieren

Olosunla Tayelolu

Biomedical uses of lithium

GRIN Verlag

Bibliografische Information der Deutschen Nationalbibliothek:

Die Deutsche Bibliothek verzeichnet diese Publikation in der Deutschen National-
bibliografie; detaillierte bibliografische Daten sind im Internet über http://dnb.d-
nb.de/ abrufbar.

Impressum:

Copyright © 2012 GRIN Verlag GmbH
Druck und Bindung: Books on Demand GmbH, Norderstedt Germany
ISBN: 978-3-656-73381-2

Dieses Buch bei GRIN:

http://www.grin.com/de/e-book/279570/biomedical-uses-of-lithium

BIOMEDICAL USES OF LITHIUM

OLOSUNLA TAYELOLU GABRIEL

A seminar submitted in partial fulfillment of the

Requirements for the award of the degree of

Master of Science (M.Sc) INORGANIC CHEMISTRY

TO

CHEMISTRY DEPARTMENT
Faculty of Science
University Of Abuja

August 2012

INTRODUCTION

Biomedical uses of Lithium refer to use of the lithium ion, Li^+, as a drug. A number of chemical salts of lithium are used medically as mood-stabilizing drugs, primarily in the treatment of bipolar_disorder, where they have a role in the treatment of depression and particularly of mania, both acutely and in the long term. As a mood stabilizer, lithium is probably more effective in preventing mania than depression, and reduces the risk of suicide in bipolar patients.(Baldessarini et al, 2006) In depression alone (unipolar disorder), lithium can be used to augment other antidepressants . Lithium carbonates ($Li2CO3$), sold under several trade names, is the most commonly prescribed, while lithium citrate ($Li3C6H5O7$) is also used in conventional pharmacological treatments. Lithium sulfate ($Li2SO4$) has been presented as an alternative[citation needed]. Lithium orotate is sometimes marketed as a "safe" natural alternative with fewer side effects than conventional lithium, yet caution must be taken with all of the active lithium salts. Upon ingestion, lithium becomes widely distributed in the central nervous system and interacts with a number of neurotransmitters and receptors, decreasing norepinephrine release and increasing serotonin synthesis.

CONCEPT OF BIOMEDICINE

Biomedicine is a branch of medical science that applies biological and other natural science principles to clinical practice. Biomedicine involves the study of (patho-) physiological processes with methods from biology, chemistry, and physics (Larson, 1998)

CHEMISTRY OF LITHIUM

Lithium is a metallic element that was discovered in 1818. Because it was found in a mineral, it was called 'lithium', which is derived from the Greek word *lithos,* stone. Lithium is identified by the symbol Li on the periodic table at the position number 3 with an atomic mass of 6.94. Lithium is used in a range of industries, typically in the form of alloys and compounds, since it is extremely reactive. Well known lithium applications include the lithium-ion battery and lithium carbonate tablet for treatment of bipolar disorder and mood stabilization(encyclopaedia britanicca,2009)

- ***Description of Lithium***

Chemical element of Group Ia in the periodic table, the alkali metal group, lightest of the solid elements. The metal itself—which is soft, white, and lustrous—and several of its alloys and compounds are produced on an industrial scale. (encyclopaedia britanicca,2009)

- ***Discovery of Lithium***

Lithium was discovered in 1817 by Johan August Arfwedson in the mineral petalite, lithium is found also in economically exploitable quantities in such minerals as spodumene, lepidolite, amblygonite, and petalite; it constitutes about 0.002 percent of the Earth's crust. Chemical treatment of the ores provides lithium hydroxide, carbonate, or sulfate, which can be converted to other compounds. Lithium metal is made by electrolyzing a molten mixture of lithium chloride and potassium chloride.The metal, which can be drawn into wire and rolled into sheets, is softer than lead but harder than the other alkali metals and has the body-centred cubic crystal structure. Lithium and its compounds impart a crimson colour to a flame, the basis of a test for its presence. Lithium floats on water, reacting with it to yield lithium hydroxide (LiOH) and hydrogen gas. It is commonly kept coated with petrolatum because it reacts with the moisture in the air. (encyclopaedia britanicca,2009)

- ***Isotopes of Lithium***

Natural lithium exists as two isotopes: lithium-7 (92.5 percent) and lithium-6 (7.5 percent); five radioactive isotopes have been prepared—lithium-5, lithium-8, lithium-9, lithium-10, and lithium-11—all having half-lives of less than one second. Lithium was used (1932) as the target metal in the pioneering work of John Cockcroft and Ernest Walton in transmuting nuclei by artificially accelerated atomic particles; each lithium nucleus that absorbed a proton became two helium nuclei. The bombardment of lithium-6 with slow neutrons produces helium and tritium. Aluminum, lead, and other soft metals can be made harder by alloying them with small proportions of lithium. (encyclopaedia britanicca,2009)

- ***Compounds of Lithiun***

Lithium is chemically active, readily losing one of its three electrons to form compounds containing the Li^+ cation. Many of these differ markedly in solubility from the corresponding compounds of the other alkali metals. A number of the lithium compounds have practical applications. Lithium hydride (LiH), a gray, crystalline solid produced by the direct combination of its constituent elements at elevated temperatures, is a ready source of hydrogen, instantly liberating that gas upon treatment with water. It also is used to produce lithium aluminum hydride (LiAlH4), which quickly reduces aldehydes, ketones, and carboxylic esters to alcohols.Lithium hydroxide (LiOH), commonly obtained by the reaction of lithium carbonate with lime, is used in making lithium salts (soaps) of stearic and other fatty acids; these soaps are widely used as thickeners in lubricating greases. Lithium hydroxide is also used as an additive in the electrolyte of alkaline storage batteries and as an absorbent for carbon dioxide. Other industrially important compounds include lithium chloride, LiCl, and lithium bromide, LiBr. They form concentrated brines capable of absorbing aerial moisture over a wide range of temperatures; these brines are commonly employed in large refrigerating and air-conditioning systems. Lithium fluoride, LiF, is used chiefly as a fluxing agent in enamels and glasses. Of greater significance is lithium carbonate, Li_2CO_3. Not only is it utilized in the preparation of other lithium compounds but it has been found to be effective in the treatment of the mental disorder manic-depressive psychosis. Organolithium compounds, in which the lithium atom is not present as the Li^+ ion but is attached directly to a carbon atom, are useful in making other organic compounds. Butyllithium, C_4H_9Li, which is used in the manufacture of synthetic rubber, is prepared by the reaction of butyl bromide, C_4H_9Br, with metallic lithium(encyclopaedia britanicca,2009)

- ***Properties of Lithium***

The atomic number is 3 and the atomic weight is 6.941The melting point is 180.5° C, boiling point is 1,342° C, specific gravity is 0.534 g/cm^3 (20° C) ,the valence is 1 and electronic configuration is . 2-1 or $1s^2 2s^1$. (encyclopaedia britanicca,2009)

HISTORICAL BACKGROUND OF BIOMEDICAL USES OF

LITHIUM

Lithium was first used in the 19th century as a treatment for gout after scientists discovered, at least in the laboratory, lithium could dissolve uric acid crystals isolated from the kidneys. The levels of lithium needed to dissolve urate in the body, however, were toxic.[Marmol,2008] Because of prevalent theories linking excess uric acid to a range of disorders, including depressive and manic disorders, Carl Lange in Denmark[Lenon and Watson, 1994] and William Alexander Hammond in New York[Mitchel, 2000] used lithium to treat mania from the 1870s onwards, though use in the form of spring waters to treat mania were reported in ancient Roman and Greek times.[citation needed] By the turn of the 20th century, this use of lithium was largely abandoned, according to Susan Greenfield, due to the reluctance of the pharmaceutical industry to invest in a drug that could not be patented.(Susan, 1999)

As accumulating knowledge indicated a role for excess sodium intake in hypertension and heart disease, lithium salts were prescribed to patients for use as a replacement for dietary table salt (sodium chloride). This practice was discontinued in 1949 when reports of side effects and deaths were published, leading to a ban of lithium sales.(Marmol,2008)

The use of lithium salts to treat mania was rediscovered by the Australian psychiatrist John Cade in 1949. Cade was injecting rodents with urine extracts taken from schizophrenic patients, in an attempt to isolate a metabolic compound which might be causing mental symptoms. Since uric acid in gout was known to be psychoactive (adenosine receptors on neurons are stimulated by it; caffeine blocks them), Cade needed soluble urate for a control. He used lithium urate, already known to be the most soluble urate compound, and observed it caused the rodents to be tranquilized. Cade traced the effect to the lithium ion itself. Soon,

Cade proposed lithium salts as tranquilizers, and soon succeeded in controlling mania in chronically hospitalized patients with them. This was one of the first successful applications of a drug to treat mental illness, and it opened the door for the development of medicines for other mental problems in the next decades.[Cade, 1949]

The rest of the world was slow to adopt this treatment, largely because of deaths which resulted from even relatively minor overdosing, including those reported from use of lithium chloride as a substitute for table salt. Largely through the research and other efforts of Denmark's Mogens Schou and Paul Baastrup in Europe,(Marmol,2008) and Samuel Gershon and Baron Shopsin in the U.S., this resistance was slowly overcome. The application of lithium in manic illness was approved by the United States Food and Drug Administration in 1970.(Mitchel,2000) In 1974, this application was extended to its use as a preventive agent for manic-depressive illness.

Lithium has become a part of Western popular culture. Characters in Pi, Premonition, Stardust Memories, American Psycho, and An Unmarried Woman all take lithium. Sirius XM Satellite Radio in North America has a 1990s alternative rock station called Lithium, and several songs refer to the use of lithium as a mood stabilizer. These include: "Lithium Lips" by Mac Lethal, "Equilibrium met Lithium" by South African artist Koos Kombuis, "Lithium" by Evanescence, "Lithium" by Nirvana, "Lithium and a Lover" by Sirenia, "Lithium Sunset", from the album Mercury Falling by Sting,(Agassi,1996) "Tea and Thorazine" by Andrew Bird, and "Lithium" by Thin White Rope.

MEDICAL USES OF LITHIUM

Lithium treatment is used to treat mania in bipolar disorder. Initially, lithium is often used in conjunction with antipsychotic drugs as it can take up to a month for it to have an effect. Lithium is also used as prophylaxis for depression and mania in bipolar disorder. It is sometimes used for other psychiatric disorders, such as cycloid psychosis and major depressive disorder.[Marsh et al, 2006][Alan et al,2007] Lithium possesses a very important antisuicidal effect not shown in other stabilizing medications such as antiseizure drugs.[Muller et al, 2003] Nonpsychiatric applications are limited; however, its use is well established in the prophylaxis of some headaches related to cluster headaches (trigeminal autonomic cephalgias), particularly hypnic headache. An Italian pilot study in humans conducted in 2005–06 suggested lithium may improve outcomes in the neurodegenerative disease amyotrophic lateral sclerosis (ALS).(Fornai, 2008) However, a randomised, double-blind, placebo-controlled trial comparing the safety and efficacy of lithium in combination with riluzole for treatment of ALS failed to demonstrate a benefit as compared to a combination therapy over riluzole alone.(Aggarwai et al, 2010)

Lithium is sometimes used as an augmenting agent to increase the benefits of standard drugs used for unipolar depression. Lithium treatment was previously considered to be unsuitable for children; however, more recent studies show its effectiveness for treatment of early-onset bipolar disorder in children as young as eight. The required dosage (15–20 mg per kg of body weight) is slightly less than the toxic level, requiring blood levels of lithium to be monitored closely during treatment. To prescribe the correct dosage, the patient's entire medical history, both physical and psychological, is sometimes taken into consideration. The starting dosage of lithium should be 400–600 mg given at night and increased weekly depending on serum monitoring. Those who use lithium should receive regular serum level tests and should monitor thyroid and kidney function for abnormalities, as it interferes with the regulation of sodium and water levels in the body, and can cause dehydration. Dehydration, which is compounded by heat, can result in increasing lithium levels. The dehydration is due to lithium inhibition of the action of antidiuretic hormone, which normally enables the kidney to reabsorb water from urine. This causes an inability to concentrate urine, leading to consequent loss of body water and thirst.(Healy,2005)   High doses of haloperidol, fluphenazine, or flupenthixol may be hazardous when used with lithium; irreversible toxic encephalopathy has been reported.(Sandyk, 1983).   Lithium salts have a narrow therapeutic/toxic ratio, so should not be prescribed unless facilities for monitoring plasma concentrations are available. Patients should be carefully selected. Doses are adjusted to achieve plasma concentrations of 0.4(Solomon et al,1996) to 1.2 mmol Li+/l (lower end of the range for maintenance therapy and elderly patients, higher end for pediatric patients) on samples taken 12 hours after the preceding dose. Overdosage, usually with plasma concentrations over 1.5 mmol Li+/l, may be fatal, and toxic effects include tremor, ataxia, dysarthria, nystagmus, renal impairment, confusion, and convulsions. If these potentially hazardous signs occur, treatment should be stopped, plasma lithium concentrations redetermined, and steps taken to reverse lithium toxicity.

Lithium toxicity is compounded by sodium depletion. Concurrent use of diuretics that inhibit the uptake of sodium by the distal tubule (e.g. thiazides) is hazardous and should be avoided. In mild cases, withdrawal of lithium and administration of generous amounts of sodium and fluid will reverse the toxicity. Plasma concentrations in excess of 2.5 mmol Li+/l are usually associated with serious toxicity requiring emergency treatment. When toxic concentrations are reached, there may be a delay of one or two days before maximum toxicity occurs.

In long-term use, therapeutic concentrations of lithium have been thought to cause histological and functional changes in the kidney. The significance of such changes is not clear, but is of sufficient concern to discourage long-term use of lithium unless it is definitely indicated. Doctors may change a bipolar patient's medication from lithium to another mood-stabilizing drug, such as valproate (Depakote), if problems with the kidneys arise. An important potential consequence of long-term lithium use is the development of renal diabetes insipidus (inability to concentrate urine). Patients should therefore be maintained on lithium treatment after three to five years only if, on assessment, benefit persists. Conventional and sustained-release tablets are available. Preparations vary widely in bioavailability, and a change in the formulation used requires the same precautions as initiation of treatment. There are few reasons to prefer any one simple salt of lithium; the carbonate has been the more widely used, but the citrate is also available.

Lithium may be used as a treatment of seborrhoeic dermatitis (lithium gluconate 8% gel). In addition, lithium has been shown to increase production of white blood cells in the bone marrow and might be indicated in patients suffering from leukopenia.(vieweg,1986)

A limited amount of evidence suggests lithium may contribute to treatment of substance abuse for some dual-disorder patients.(Frye and Saloume, 2006)

In 2009, Japanese researchers at Oita University reported low levels of naturally occurring lithium in drinking water supplies reduced suicide rates.[] A previous report had found similar data in the American state of Texas.(Gonzalenz et al, 2008) In response, psychiatrist Peter Kramer raised the hypothetical possibility of adding lithium to drinking water as a mineral supplement rather than as a therapeutic drug.(The therapeutic dosage of lithium carbonate (tablets and capsules) or citrate (liquid) "usually ranges from 900 - 1,200 mg/day" and is adjusted according to patient response and blood levels.) This is analogous to niacin, where a low dose in multivitamin pills is taken as a vitamin supplement to prevent the niacin deficiency disease pellagra, but a high dose is prescribed as a therapeutic drug to raise high-density lipoprotein ("good" cholesterol) levels.

Lithium was the first specific psychotropic medication, predating the neuroleptics by several years (Gelenberg et al, 1995) and the antidepressants by almost a decade. According to Goodwin & Ghaemi (Werder,1995),it heralded the ''psychopharmacological revolution''. The impact of Cade's discovery can also be considered at many other levels: the relief of suffering for multitudinous bipolar patients and their families; the economic benefits to the wider community (it has been estimated that from 1970 to 1994 lithium saved the USA alone over US$ 145 billion dollars in hospitalization costs (Larson,1995); the solid underpinning of Kraepelin's distinction between dementia praecox (schizophrenia) and manic-depressive insanity (bipolar disorder); and a resurgence of the interest in the biological roots of the ''functional'' psychoses that had been largely lost since the 19th century.

Psychiatrists use lithium as a therapeutic agent mainly for treating the following disorders:

1. Bipolar (manic depression) disorder

2. As mood stabilizing agent

3. Schizophrenia and Alzheimer's Disease

BIPOLAR DISORDER

It is formerly called *manic depression or manic-depressive illness* . It is a mental disorder characterized by severe and recurrent depression or mania with abrupt or gradual onsets and recoveries.(Thompson, 2002) The states of mania and depression may alternate cyclically, one mood state may predominate over the other, or they may be mixed or combined with each other. A bipolar person in the depressive phase may be sad, despondent, listless, lacking in energy, and unable to show interest in his surroundings or to enjoy himself and may have a poor appetite and disturbed sleep(Chao et al, 2005). The depressive state can be agitated—in which case sustained tension, overactivity, despair, and apprehensive delusions predominate—or it can be retarded—in which case the person's activity is slowed and reduced, he is sad and dejected, and he suffers from self-depreciatory and self-condemnatory tendencies. Mania is a mood disturbance that is characterized by abnormally intense excitement, elation, expansiveness, boisterousness, talkativeness, distractibility, and irritability. The manic person talks loudly, rapidly, and continuously and progresses rapidly from one topic to another; is extremely enthusiastic, optimistic, and confident; is highly sociable and gregarious; gesticulates and moves about almost continuously; is easily irritated and easily distracted; is prone to grandiose notions; and shows an inflated sense of self-esteem. The most extreme manifestations of these two mood disturbances are, in the manic phase, violence against others and, in the depressive, suicide. A bipolar disorder may also feature such psychotic symptoms as delusions and hallucinations. (Bertholf et al,1998)Depression is the more common symptom, and many patients never develop a genuine manic phase, although they may experience a brief period of over optimism and mild euphoria while recovering from a depression.(Gelenberg et al, 1995)

Bipolar disorders of varying severity affect about 1 percent of the general population and account for 10 to 15 percent of readmissions to mental institutions. Statistical studies have suggested a hereditary predisposition to bipolar disorder, and this predisposition has now been linked to a defect on a dominant gene located on chromosome 11. In a physiological sense, it is believed that bipolar disorder is caused by the faulty regulation of one or more naturally occurring amines at sites in the brain where the transmission of nerve impulses takes place; a deficiency of the amines results in depression, and an excess of them causes mania. The most likely candidates for the suspect amines are norepinephrine, dopamine, and 5-hydroxytryptamine. The ingestion of lithium carbonate on a long-term basis has been found effective in alleviating or even eliminating the symptoms of many persons with bipolar disorder. Bipolar disorder was described in antiquity by the 2nd-century Greek physician Aretaeus of Cappadocia and definitively in modern times by the German psychiatrist Emil Kraepelin.(Chao et al,2005)

- ***Causes of bipolar disorder***

While the exact cause of bipolar disorder is not entirely known, genetic, neurochemical and environmental factors probably interact at many levels to play a role in the onset and progression of bipolar disorder. There is a great deal of scientific evidence that indicates bipolar disorder is due to a chemical imbalance in the brain, or a malfunction of the neurotransmitter glutamate in the brain.(Werder, 1995) Excessive amounts of glutamate in the space between neurons causes mania, and too little, depression.(Thompson, 2002)

- ***People affected by bipolar disorder***

World Wide: 222 million, according to AstraZeneca's bipolar statistics: "Between 3 and 4% of the world's adult population is affected by bipolar disorder. That is 222 million adults worldwide."The United States: 6 million, according to the latest National Institute of Mental Health, "Bipolar disorder affects approximately 5.7 million adult Americans, or about 2.6% of the U.S. population age 18 and older every year." China and India: 15 and 12 million respectively, when using 1.2% of the population as the prevalence rate for extrapolation. Bipolar disorder statistics from the World Health Organization (WHO), indicate that bipolar disorder is the 6th leading cause of disability in the world.

- ***The way lithium work on bipolar disorder***

Studies indicate that lithium keeps the quantity of the neurotransmitter glutamate stable and at optimal levels. Dr. Lowell Hokin, a University of Wisconsin professor of pharmacology, theorized that lithium works to control bipolar because of its stabilizing effect on glutamate receptors, exerting a dual effect on receptors for the neurotransmitter glutamate, acting to keep the amount of glutamate active between cells at a stable, healthy level, neither too much nor too little.

- ***Effectiveness of lithium therapy***

Despite the potential difficulties with lithium treatment, lithium remains the best medication for treating bipolar disorder and for stabilizing mood in most people. Even after 35 years since it was approved by the FDA, lithium is still the gold standard and first choice for treatment of bipolar disorder and mood stabilization.

Lithium is successful in improving both the manic and depressive symptoms in up to as many as 70 to 80% of patients. It is still considered the drug of choice for reduction of suicide risk in bipolar patients. Lithium therapy can lead to fewer, less severe, and shorter manic or hypomanic episodes. Many people find lithium an effective medicine that helps to control their mood disorder and greatly improve their quality of life.

- ***Admnistration of lithium***

Lithium carbonate (Li2CO3) is the most commonly used pharmaceutical form of lithium. It is formulated in tablet form and administrated orally.

Lithium tablets are normally taken as a single dose at night as this is more convenient and re-duces the problems with some of the side effects. Lithium is usually taken for one to two years to derive full benefit from its use. Many people need to stay on it long-term to prevent the illness from relapsing. When initiating lithium therapy, dosage must be individualized according to serum levels and clinical responses.

For Acute Mania: Optimal patient response can usually be established and maintained with 1,800 mg per day. Such doses will normally produce the desired serum lithium level ranging between 1.0 and 1.5 mM.

For Long-Term Control: The desirable serum lithium levels are 0.6 to 1.2 mM. Dosage will vary from one individual to another, but usually 900 mg to 1,200 mg per day will maintain this level. Serum lithium levels in uncomplicated cases receiving maintenance therapy during remission should be monitored at least every two months.

- *The side effects of lithium therapy*

The most common side effects are an overall dazed feeling and a fine hand tremor. These side effects are generally present during the length of the treatment, but can sometimes disappear in certain patients. Other common side effects, such as nausea and headache, can be generally remedied by a higher intake of water. Lithium unbalances electrolytes; to counteract this, increased water intake is suggested.

Lithium is known to be responsible for significant amounts of weight gain.Barbara, 2001) Because lithium competes with the receptors for the antidiuretic hormone in the kidney, it increases water output into the urine, a condition called nephrogenic diabetes insipidus. Clearance of lithium by the kidneys is usually successful with certain diuretic medications, including amiloride and triamterene.(Wetzels, 1989) It increases the appetite and thirst ("polydypsia") and reduces the activity of thyroid hormone (hypothyroidism).(Keshava, 2001) It is also believed to affect renal function (GFR).(Bendz,2010)

Lithium is a well-known cause of downbeat nystagmus,(Lee, 2003) which may be permanent or require several months of abstinence for improvement.(Williams,1998)

Teratogenicity

Lithium is also a teratogen, causing birth defects in a small number of newborn babies.(Shepherd et al, 2002) Case reports and several retrospective studies have demonstrated possible increases in the rate of a congenital heart defect known as Ebstein's anomaly, if taken during a woman's pregnancy.(Yacobi, 2008) As a consequence, fetal echocardiography is routinely performed in pregnant women taking lithium to exclude the possibility of cardiac anomalies. Lamotrigine seems to be a possible alternative to lithium in pregnant women.Yacobi, 2008) Gabapentin (Montouris, 2003) and clonazepam(Weinstock et al, 2001) are also indicated as antipanic medications during the childbearing years and during pregnancy. Valproic acid and carbamazepine also tend to be associated with teratogenicity.

The average developmental score for the children exposed to lithium *in utero* was 7–8 points lower than the control group (siblings), but well within the normal range of 100±15. Cognitive depression in adults is disputed.

Dehydration

Dehydration in patients taking lithium salts can be very hazardous, especially when combined with lithium induced nephrogenic diabetes insipidus with polyuria. Such situations include preoperative fluid regimen or other fluid inaccessibility, warm water conditions, sporting events, and hiking. Another dangerous situation is when low sodium only hydration may very quickly produce hyponatremia with its danger of toxic lithium concentrations in plasma.

Overdose

Lithium toxicity may occur in persons taking excessive amounts either accidentally or intentionally on an acute basis or in patients who accumulate high levels during ongoing chronic therapy. The manifestations include nausea, emesis, diarrhea, asthenia, ataxia, confusion, lethargy, polyuria, seizures and coma. Other toxic effects of lithium include coarse tremor, muscle twitching, convulsions and renal failure.(Gelder, 2005) Persons who survive a poisoning episode may develop persistent neurotoxicity.

MECHANISM OF ACTION OF LITHIUM

Unlike other psychoactive drugs, Li+ typically produces no obvious psychotropic effects (such as euphoria) in normal individuals at therapeutic concentrations. Li+ possibly produces its effects by interacting with the transport of monovalent or divalent cations in neurons. However, because it is a poor substrate at the sodium pump, it cannot maintain a membrane potential and only sustains a small gradient across biological membranes.[citation needed] Li+ is similar enough to Na+ that under experimental conditions, it can replace Na+ for production of a single action potential in neurons.

Recent research suggests three different mechanisms which may or may not act together to deliver the mood-stabilizing effect of this ion.(Amdisen,1978) The excitatory neurotransmitter glutamate could be involved in the effect of lithium as other mood stabilizers, such as valproate and lamotrigine, exert influence over glutamate, suggesting a possible biological explanation for mania.[citation needed] The other mechanisms by which lithium might help to regulate mood include the alteration of gene expression.(Baselt, 2008)

An unrelated mechanism of action has been proposed in which lithium deactivates the GSK3β enzyme.(Jope, 1999) This enzyme normally phosphorylates the Rev-Erbα transcription factor protein stabilizing it against degradation. Rev-Erbα in turn represses BMAL1, a component of the circadian clock. Hence, lithium by inhibiting GSK3β causes the degradation of Rev-Erbα and increases the expression of BMAL which dampens the circadian clock(Lenox, 2003) Through this mechanism, lithium is able to block the resetting of the "master clock" inside the brain; as a result, the body's natural cycle is disrupted. When the cycle is disrupted, the routine schedules of many functions (metabolism, sleep, body temperature) are disturbed. Lithium may thus restore normal brain function after it is disrupted in some people.

Another mechanism proposed in 2007 is that lithium may interact with nitric oxide (NO) signalling pathway in the central nervous system, which plays a crucial role in the neural plasticity. The NO system could be involved in the antidepressant effect of lithium in the Porsolt forced swimming test in mice.(Weinstock et al, 2007) It was also reported that NMDA receptor blockage augments antidepressant-like effects of lithium in the mouse forced swimming test,(Yin et al,2006) indicating the possible involvement of NMDA receptor/NO signaling in the action of lithium in this animal model of learned helplessness. Lithium treatment has been found to inhibit the enzyme inositol monophosphatase, leading to higher levels of inositol triphosphate.(Gbasemi et al, 2009) This effect was enhanced further with an inositol triphosphate reuptake inhibitor. Inositol disruptions have been linked to memory impairment and depression.

- ***Why and when is lithium blood levels tested?***

Lithium has a narrow therapeutic range (0.4-1.4 mM), and too low of a dosage leads to ineffectiveness and too high leads to severe toxicity. Therefore regular monitoring of the patient's clinical state and serum lithium levels is required to:

(1) identify and /or prevent potential toxicity associated with high levels

(2) assure ongoing efficacy and effectiveness

(3) monitor the patient's adherence to the prescribed regimen

The lithium test may be ordered frequently (every few days) when a patient first begins taking lithium or if a patient is returning to its use after an absence. This is done to help adjust the dose to the desired blood level. The test may be ordered at regular intervals or as needed to monitor blood concentrations. One or more lithium tests may be ordered if a patient starts taking additional medications (to judge their effect, if any, on lithium levels) and may be ordered if the doctor suspects toxicity.

Once stable blood concentrations in the therapeutic range have been achieved, lithium may then be monitored at regular intervals to ensure that it remains in this range. The test may also be ordered when a patient's condition does not appear to be responding to initial lithium dosage levels in order to determine whether concentrations are too low, the medication is ineffective, and/or to determine if the patient is complying with therapy (taking the lithium regularly). It may also be ordered when a patient experiences a troublesome level of side effects and/or exhibits symptoms that the doctor suspects may be due to toxicity.

Patients should talk to their doctor about the timing of the sample collection. Lithium blood levels are generally performed 12-18 hours after the last dose. Since dosage timing varies and some formulations are time released, collection specifics may vary.

METHODS USED FOR LITHIUM TESTING

Methods used for lithium testing in clinical laboratories are continuously evolving. Before 1987, lithium was measured by flame atomic emission spectrometry (FAES) in about 90% of laboratories and by flame atomic absorption spectrometry (FAAS) in the rest. In 1987, the first generation Ion-Selective Electrode (ISE) technique based lithium analyzer (NOVA Biomedical, Co. Waltham, MA) was introduced, and within 2 years, (by 1989), the ISE lithium test reached approximately 20% of the total clinical lithium testing, owing to its cost advantages and ease of use in comparison to the FAES method. In 2001, a colorimetric assay was developed by Trace American Inc. (now Thermo Fisher) utilizing a reaction between lithium and a porphyrin dye. This dye based lithium assay immediately became popular in clinical laboratories as it can be used in general chemistry analyzers and does not require a specially dedicated instrument, even though it is a relatively expensive test. In 2008, a new colorimetric, liquid stable enzymatic lithium test was introduced by Diazyme Laboratories. The enzymatic lithium test utilizes a recombinant enzyme that is highly sensitive to the inhibitory effect of lithium. This novel lithium test offers several key advantages over older lithium test methods especially in assay accuracy, reagent and calibration curve stability, user friendliness, and as cost effectiveness. The enzymatic lithium assay is expected to become the first of choice for clinical laboratories around the world.

A brief introduction of lithium test methods is depicted below:

1. Flame Atomic Emission Spectrometry (FAES) method

Flame atomic emission spectrometry (FAES) uses quantitative measurement of the optical emission from excited atoms to determine analyte concentration. Analyte atoms in solution are aspirated into the excitation region where they are desolvated, vaporized, and atomized by a flame, discharge, or plasma as shown in the scheme below.

This method is not automated, and often requires sample preparations, and is a costly method.

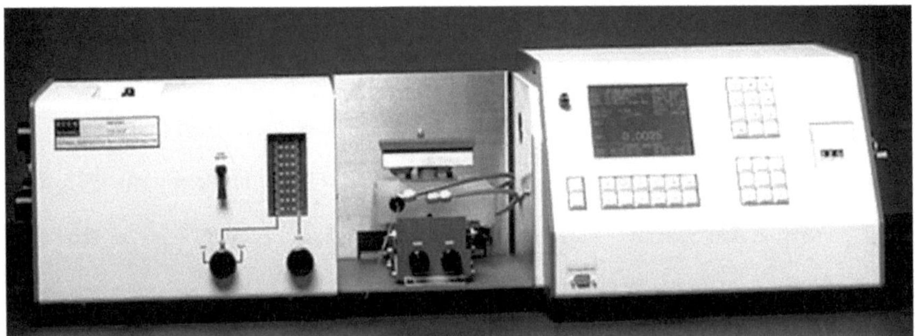

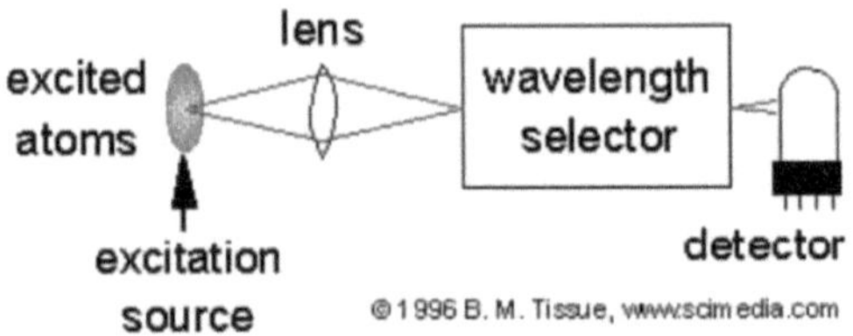

2. Ion-Selective Electrode (ISE) method

Ion-Selective-Electrode (ISE) methods utilize techniques which involve crown ethers having a core which accommodate the Li+ ion. While early ISE techniques were prone to interferences caused by sodium and other ions present in the test sample, more recent ISE analyzers are capable of accurate and precise lithium measurement. Although ISE analyzers are less expensive and easier to operate than flame photometer equipment, the electrode probe maintenance is costly, and requires dedicated instrumentation and trained personnel.

3. Porphyrin Dye Spectrometry Method

The porphyrin dye based lithium assay employs the dye porphyrin which reacts with Li in an alkaline condition to form a noncovalent binary complex, resulting in a change in absorbance at 510 nm, which is in direct proportion to Li concentration in the sample. This method is fully automated and can be used in most of chemistry analyzers. However, there are some limitations for this method. These include:

- highly sensitive to light

- extremely high pH (pH >12) of the r eagent

- short on-board reagent stability and calibration curve stability

- requires sample preparation (dilution)

- reagent is corrosive to instruments and hazardous to ship

- absorbs atmospheric carbon dioxide (CO_2), resulting in poor precision.

- high cost per test

4. Liquid Stable Enzymatic Method

The Diazyme Enzymatic Lithium Assay is based on a lithium sensitive enzyme whose activity is lithium concentration dependent. The enzyme, a phosphatase, converts its substrate adenosine biphosphate (PAP) to hypoxanthine through a coupled enzymatic reaction to generate hydrogen peroxide ($H2O2$), which is quantified by the conventional Trinder reaction. The lithium concentration in the sample is inversely proportional to the enzyme activity or the amount of $H2O2$ produced in the reaction.

REFERENCES

- Lewis, R. (1996). Evening Out the Ups and Downs of Manic Depressive Illness. FDA Consumer, 30 (5), 26-29.

- Gelenberg, AJ, Hirschfeld, RMA, Jefferson, JW, Potter, WZ, Thase, ME. (1995). Managing Lithium Treatment. Patient Care, 29 (19), 71-86

- Werder, SF (1995). An Update on the Diagnosis and Treatment of Mania in Bipolar Disorder. American Family Physican, 51 (5) 1126-1136.

- Larson, R. (1998) Lithium Prevents Sucides. Insights on the News, 14 (18), 39.

- More Uses for "Miracle Drug," Lithium Sought in its 50th Year. (1999). Psychopharmacology Update, 10 (7), 1.

- Bertholf RL et al. (1988) Lithium determined in serum with an ion-selective electrode. Clin Chem. 34: 1500-2

- Chapoteau E. et al. (1992) First practical colorimetric assay of lithium in serum. Clin. Chem. 36: 1654-7

- Thompson JC. (2002) Development of an automated photometric assay for lithium on the dimension clinical chemistry system. Clin. Chem. Acta. 327: 149-56

- Chao Dou at al. (2005). Automated enzymatic assay for measurement of lithium ions in human serum. Clin. Chem. 51: 1989-1991

- Baldessarini, Ross J; Tondo, Leonardo; Davis, Paula; Pompili, Maurizio; Goodwin, Frederick K; Hennen, John (2006). "Decreased risk of suicides and attempts during long-term lithium treatment: A meta-analytic review". Bipolar Disorders 8 (5p2): 625–39. doi:10.1111/j.1399-5618.2006.00344.x. PMID 17042835.

- Mash, Eric J.; Russell A. Barkley (2006). Treatment of childhood disorders. Guilford Press. p. 443. ISBN 1-57230-921-0. http://books.google.com/?id=KCZqs_NlfmQC.

- Schatzberg, Alan F.; Jonathan O. Cole, Charles DeBattista (2007). Manual of clinical psychopharmacology. American Psychiatric Publishing. p. 267. ISBN 1-58562-317-2. http://books.google.com/?id=NvT0l6iL5IQC.

- Müller-Oerlinghausen, B; Berghöfer, A; Ahrens, B (2003). "The Antisuicidal and Mortality-Reducing Effect of Lithium Prophylaxis: Consequences for Guidelines in Clinical Psychiatry". Canadian Journal of Psychiatry 48 (7): 433–9. PMID 12971012. https://ww1.cpa-apc.org/Publications/Archives/CJP/2003/august/muller.asp.

- Kovacsics, Colleen E.; Gottesman, Irving I.; Gould, Todd D. (2009). "Lithium's Antisuicidal Efficacy: Elucidation of Neurobiological Targets Using Endophenotype Strategies". Annual Review of Pharmacology and Toxicology 49: 175–198. doi:10.1146/annurev.pharmtox.011008.145557. PMID 18834309.

- Fornai, F.; Longone, P.; Cafaro, L.; Kastsiuchenka, O.; Ferrucci, M.; Manca, M. L.; Lazzeri, G.; Spalloni, A. et al. (2008). "Lithium delays progression of amyotrophic lateral sclerosis". Proceedings of the National Academy of Sciences 105 (6): 2052. doi:10.1073/pnas.0708022105.

- Aggarwal, Swati P; Zinman, Lorne; Simpson, Elizabeth; McKinley, Jane; Jackson, Katherine E; Pinto, Hanika; Kaufman, Petra; Conwit, Robin A et al. (2010). "Safety and efficacy of lithium in combination with riluzole for treatment of amyotrophic lateral sclerosis: A randomised, double-blind, placebo-controlled trial". The Lancet Neurology 9 (5): 481. doi:10.1016/S1474-4422(10)70068-5.

- Healy D. 2005. Psychiatric Drugs Explained. 4th ed. Churchhill Livingstone: London.[page needed]

- (Sandyk, R.; Hurwitz, M. D. (1983). "Toxic irreversible encephalopathy induced by lithium carbonate and haloperidol. A report of 2 cases". South African medical journal = Suid-Afrikaanse tydskrif vir geneeskunde 64 (22): 875–876.

- Gille, M.; Ghariani, S.; Piéret, F.; Delbecq, J.; Depré, A.; Saussu, F.; De Barsy, T. (1997). "Acute encephalomyopathy and persistent cerebellar syndrome after lithium salt and haloperidol poisoning". Revue neurologique 153 (4): 268–270. PMID 9296146. edit)

- ^ One study (Solomon, D.; Ristow, W.; Keller, M.; Kane, J.; Gelenberg, A.; Rosenbaum, J.; Warshaw, M. (1996). "Serum lithium levels and psychosocial function in patients with bipolar I disorder". The American Journal of Psychiatry 153 (10): 1301–1307. PMID 8831438. edit) concluded a "low" dose of 0.4–0.6 mmol/L serum lithium treatment for patients with bipolar 1 disorder had less side effects, but a higher rate of relapse, than a "standard" dose of 0.8–1.0 mmol/l. However, a reanalysis of the same experimental data

- (Perlis, R.; Sachs, G.; Lafer, B.; Otto, M.; Faraone, S.; Kane, J.; Rosenbaum, J. (2002). "Effect of abrupt change from standard to low serum levels of lithium: A reanalysis of double-blind lithium maintenance data". The American Journal of Psychiatry 159 (7): 1155–1159. PMID 12091193. edit) concluded the higher rate of relapse for the "low" dose was due to abrupt changes in the lithium serum levels[improper synthesis?]

- Vieweg, W. V. R.; Yank, Rowe, Hovermale, Clayton (Fall 1986). "Increase in White Blood Cell Count and Serum Sodium Level Following the Addition of Lithium to Carbamazephine Treatment among three chronically Psychotic male Patients with disturbed Affective States". Psychiatric Quarterly 583: 213. http://resources.metapress.com/pdf-preview.axd?code=u11w152782481762&size=largest. Retrieved 2010-04-20.

- Rosenberg, J.; Salzman, C. (2007). "Update: New uses for lithium and anticonvulsants". CNS spectrums 12 (11): 831–841. PMID 17984856. edit

- Frye, M. A.; Salloum, I. M. (2006). "Bipolar disorder and comorbid alcoholism: Prevalence rate and treatment considerations". Bipolar Disorders 8 (6): 677–685. doi:10.1111/j.1399-5618.2006.00370.x. PMID 17156154. edit

- Vornik, L.; Brown, E. (2006). "Management of comorbid bipolar disorder and substance abuse". The Journal of clinical psychiatry 67 Suppl 7: 24–30. PMID 16961421. edit

- Ohgami, H; Terao, T; Shiotsuki, I; Ishii, N; Iwata, N (2009). "Lithium levels in drinking water and risk of suicide". The British journal of psychiatry : the journal of mental science 194 (5): 464–5; discussion 446. doi:10.1192/bjp.bp.108.055798. PMID 19407280.

- Gonzalez, R; Bernstein, I; Suppes, T (2008). "An investigation of water lithium concentrations and rates of violent acts in 11 Texas counties: Can an association be easily shown?". The Journal of clinical psychiatry 69 (2): 325–6. PMID 18363457.

- Sperner-Unterweger, Barbara; W. Wolfgang Fleischhacker, Wolfgang P. Kaschka (2001). Psychoneuroimmunology. Karger Publishers. p. 22. ISBN 3-8055-7262-X. http://books.google.com/?id=Q0xJfkY5VRUC.

- Wetzels, JF; Van Bergeijk, JD; Hoitsma, AJ; Huysmans, FT; Koene, RA (1989). "Triamterene increases lithium excretion in healthy subjects: Evidence for lithium transport in the cortical collecting tubule". Nephrology, dialysis, transplantation : official publication of the European Dialysis and Transplant Association - European Renal Association 4 (11): 939–42. PMID 2516883.

- Keshavan, Matcheri S.; John S. Kennedy (2001). Drug-induced dysfunction in psychiatry. Taylor & Francis. p. 305. ISBN 0-89116-961-X. http://books.google.com/?id=pol0204fqjIC.

- Bendz, Hans; Schön, Attman, Aurell (February 1, 2010). "Renal failure occurs in chronic lithium treatment but is uncommon". Kidney International 77 (3): 219–24. doi:10.1038/ki.2009.433. PMID 19940841.

- Lee, Michael S.; Lessell, Simmons (January 28, 2003). "Lithium-induced periodic alternating nystagmus". Neurology (journal) 60 (2): 344. doi:10.1212/01.WNL.0000042787.51461.D1. PMID 12552061.

- Williams, Douglas P.; Jack Rogers and B. Todd Troost (September 1988). "Lithium-Induced Downbeat Nystagmus". Archives of Neurology 45 (9): 1022–1023. doi:10.1001/archneur.1988.00520330112019. PMID 3137915.

- Shepard, TH.; Brent, RL.; Friedman, JM.; Jones, KL.; Miller, RK.; Moore, CA.; Polifka, JE. (2002). "Update on new developments in the study of human teratogens". Teratology 65 (4): 153–61. doi:10.1002/tera.10032. PMID 11948561.

- Yacobi, S; Ornoy, A (2008). "Is lithium a real teratogen? What can we conclude from the prospective versus retrospective studies? A review". The Israel journal of psychiatry and related sciences 45 (2): 95–106. PMID 18982835.

- Montouris, G (2003). "Gabapentin exposure in human pregnancy: Results from the Gabapentin Pregnancy Registry". Epilepsy & behavior : E&B 4 (3): 310–7. doi:10.1016/S1525-5050(03)00110-0. PMID 12791334.

- Weinstock, L; Cohen, LS; Bailey, JW; Blatman, R; Rosenbaum, JF (2001). "Obstetrical and neonatal outcome following clonazepam use during pregnancy: A case series". Psychotherapy and psychosomatics 70 (3): 158–62. PMID 11340418.

- Gelder, M., Mayou, R. and Geddes, J. 2005. Psychiatry. 3rd ed. New York: Oxford. pp249.

- Amdisen A. (1978). "Clinical and serum level monitoring in lithium therapy and lithium intoxication". J. Anal. Toxicol. 2: 193–202.

- R. Baselt, Disposition of Toxic Drugs and Chemicals in Man, 8th edition, Biomedical Publications, Foster City, CA, 2008, pp. 851–854.

- Jope RS (1999). "Anti-bipolar therapy: mechanism of action of lithium". Mol. Psychiatry 4 (2): 117–128. doi:10.1038/sj.mp.4000494. PMID 10208444.

- Lenox RH, Wang L (February 2003). "Molecular basis of lithium action: integration of lithium-responsive signaling and gene expression networks". Mol. Psychiatry 8 (2): 135–44. doi:10.1038/sj.mp.4001306. PMID 12610644.

- Klein PS, Melton DA (August 1996). "A molecular mechanism for the effect of lithium on development". Proc. Natl. Acad. Sci. U.S.A. 93 (16): 8455–9. doi:10.1073/pnas.93.16.8455. PMC 38692. PMID 8710892. http://www.pubmedcentral.nih.gov/articlerender.fcgi?tool=pmcentrez&artid=38692.

- Yin L, Wang J, Klein PS, Lazar MA (February 2006). "Nuclear receptor Rev-erbalpha is a critical lithium-sensitive component of the circadian clock". Science 311 (5763): 1002–5. doi:10.1126/science.1121613. PMID 16484495. Lay summary – The Scientist.

- Ghasemi M, Sadeghipour H, Mosleh A, Sadeghipour HR, Mani AR, Dehpour AR (May 2008). "Nitric oxide involvement in the antidepressant-like effects of acute lithium administration in the mouse forced swimming test". Eur Neuropsychopharmacol 18 (5): 323–32. doi:10.1016/j.euroneuro.2007.07.011. PMID 17728109.

- Ghasemi M, Sadeghipour H, Poorheidari G, Dehpour AR (June 2009). "A role for nitrergic system in the antidepressant-like effects of chronic lithium treatment in the mouse forced swimming test". Behav. Brain Res. 200 (1): 76–82. doi:10.1016/j.bbr.2008.12.032. PMID 19166880.

- Ghasemi M, Raza M, Dehpour AR (April 2010). "NMDA receptor antagonists augment antidepressant-like effects of lithium in the mouse forced swimming test". J. Psychopharmacol. (Oxford) 24 (4): 585–94. doi:10.1177/0269881109104845. PMID 19351802.

- Einat H, Kofman O, Itkin O, Lewitan RJ, Belmaker RH (1998). "Augmentation of lithium's behavioral effect by inositol uptake inhibitors". J Neural Transm 105 (1): 31–8. doi:10.1007/s007020050035. PMID 9588758.

- Marmol, F. (2008). "Lithium: bipolar disorder and neurodegenerative diseases Possible cellular mechanisms of the therapeutic effects of lithium". Progress in neuro-psychopharmacology & biological psychiatry 32 (8): 1761–1771. doi:10.1016/j.pnpbp.2008.08.012. PMID 18789369. edit

- Lenox, RH; Watson, DG (1994). "Lithium and the brain: a psychopharmacological strategy to a molecular basis for manic depressive illness". Clinical chemistry 40 (2): 309–14. PMID 8313612.

- Mitchell, PB; Hadzi-Pavlovic, D (2000). "Lithium treatment for bipolar disorder". Bulletin of the World Health Organization 78 (4): 515–7. PMC 2560742. PMID 10885179. https://www.who.int/entity/bulletin/archives/78(4)515.pdf.

- Greenfield, Susan: "Brain Power: Working out the Human Mind", page 91. Element Books Limited, 1999

- Cade J. F. J. (1949). "Lithium salts in the treatment of psychotic excitement" (PDF). Medical Journal of Australia 2 (10dfbnm): 349–52. PMID 18142718. http://www.who.int/docstore/bulletin/pdf/2000/issue4/classics.pdf.

- P. B. Mitchell, D. Hadzi-Pavlovic (2000). "Lithium treatment for bipolar disorder" (PDF). Bulletin of the World Health Organization 78 (4): 515–7. PMC 2560742. PMID 10885179. http://www.who.int/docstore/bulletin/pdf/2000/issue4/classics.pdf.